MYSTERIOUS WORLD

THE FUTURE

Investigations into the unknown

Ivor Baddiel & Tracey Blezard

MACDONALD YOUNG BOOKS

Contents

About this book

Could people spend a weekend on Mars one day? Might a race of human robots take over our lives? Will we finally find a way to cheat death forever?

Predicting the future has never been easy. A hundred years ago, the idea of Moon landings was absurd. In 1899, Charles Durrel at the US patents office declared: 'Everything that can be invented has been invented.' And in 1943, the

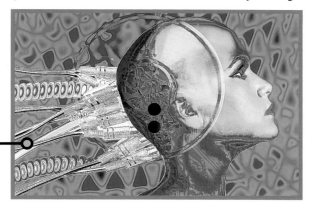

chairman of IBM, who manufactured business machines, thought the world market for computers would only number around five!

So what surprises will the 21st century hold? Here is your chance to make a forecast for the future. Look at the predictions, consider what science says today, and then make up your own mind. Science fiction or future fact? Fantasy or reality?

No one can know for sure what lies ahead. Unforeseen events might happen tomorrow and change all our futures beyond anything we can imagine. There could be definite proof of life in other galaxies, or a comet that crashes to Earth destroying everything in its path. All we can do is look forward to the strange world of tomorrow and continue to expect the unexpected!

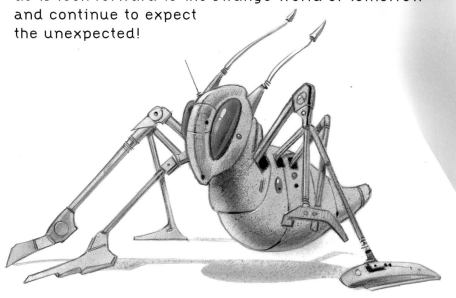

Lifestyles

Fifty years ago, futurologists believed the year 2000 would bring endless leisure time and a life of peace and harmony.

? THE MYSTERY

What changes can we expect to our lifestyles in the future?

THE PREDICTIONS

Nanotechnology will alter everything. This involves manipulating the individual atoms of which everything is made. Everyday objects will become 'smart', changing automatically to suit your needs. At home, chairs will mould to your body and tables will adjust to your height.

Everything will be miniaturized. Computers will be strapped to your wrist and incorporate a watch and mobile phone.

Computer-generated virtual reality will replace real activities. People will take virtual holidays and work out on virtual exercise machines. In school, children will play on virtual sports fields and playground games will be replaced by hi-tech, computer-generated versions.

A single nano-gadget will prepare instant food from household waste, such as reconstituted cardboard 'beefburgers'. A holographic chef will offer menu suggestions.

Eating animals will no longer be necessary or acceptable and hunting will be illegal. Zoos will disappear and the animals returned to natural environments. Rare creatures will be cloned to prevent them dying out.

Schools will operate shift systems with children attending morning or afternoon sessions. Twice as many pupils can then be educated on the same premises.

All but a handful of libraries will close as more books are transmitted electronically. The latest information will be accessible via the Internet.

To relax, you will have a home entertainment unit combining TV, cinema, computer, games console and hi-fi. Instead of a fixed screen, holographic images will play in the air.

Technology will rule people's lives. School tests, job interviews, dating, getting married or divorced, will all be done via voice-activated computers.

Cash will disappear from circulation. Everyone will carry just one 'smart card'.

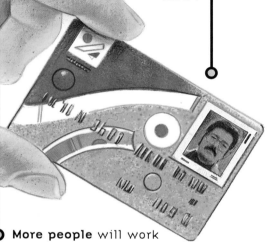

More people will work from home, hooked up to the office via an 'Intranet' – a mini Internet system.

With computers and machines taking over our lives, we will forget how to think for ourselves.

Nations will still go to war. Insect-sized robots will sabotage armies by climbing unseen into machines.

The gap between rich and poor, developing and developed countries, will widen. The rich will live in communities patrolled by guards. An 'info-poor' class will emerge, unable to access the Internet.

WHAT SCIENCE SAYS

Nanotechnology Small is the key to future technology. German scientists recently built and flew a 2 cm-long helicopter. Engines the size of a sugar grain, with gears smaller than a human hair, have been developed in New Mexico. Today's smallest computer attaches to your belt. By manipulating the atoms of a structure, 'thinking' homes that adapt to your needs are a real possibility.

Virtual reality Virtual reality simulators are used in many areas of our lives: soldiers train on them; architects design on them; many of us play games on them. Future uses will be even more advanced.

Food Futurologists predict less animal protein in our diets by 2025. Nano food machines could produce meals from scraps by breaking them down into atoms and reforming them.

Computers Shopping and banking via computer is already possible. In one US state, computers now grant instant divorces. Voice-activated computers should be common in 5 years.

Home Entertainment A Korean company is developing 3D TV. Holographic film sequences have been produced but so far only in red!

Cash Electronic money (where transactions are shown on computer) will eventually replace cash and credit cards. The idea of a multi-purpose 'smart card', pioneered in France, has been tested in Britain.

Work By 2025, 40% of people will work from home. A 'digital desk' with an internal computer controlled by hand movements already exists.

Rich and poor Gated communities guarded by CCTV are springing up in some parts of the US and Britain.

Technology is accelerating but 60% of the world's population have yet to use a telephone!

War Robot 'soldier ants' are currently being developed, observed by an interested US army.

What Do You Think?

Who will look after our electronic money?

If scientists can alter the world at an atomic level, will everything change?

Can virtual reality sometimes be better than actual reality?

Could life ever become too easy?

Is it wrong to eat animals?

Getting Around

THE MYSTERY

How far and how fast will future transportation devices take us?

In 1903, when the Wright brothers' flying machine covered 38 km in 38 minutes, they were ecstatic. Imagine their reaction to the *Aurora* spyplane's speed of 8,495 km per hour in the 1990s.

THE PREDICTIONS

Running on alternative energy sources, cars will be environmentally friendly. They will be amphibious and computer-controlled for comfort and safety. Road tolls, pedestrianized cities with moving walkways and efficient public transport will also reduce pollution.

Solar-powered planes will be developed that can fly for months at a time at a fraction of current costs.

Space rockets will be powered by highly explosive, subatomic antimatter or lasers, using 20 times less fuel than now.

Re-usable rockets will make holidays in space possible. Eventually, self-sufficient communities will be established on the Moon and Mars.

Pilotless, saucer-shaped planes will zip into orbit en route to their final destinations. A trip from London to New York (via space) would take just 30 minutes.

For those wishing to stay on this planet, a weekend break anywhere in the world will be easy and affordable. Virtual-reality holidays from your armchair will be even simpler!

High-speed trains will hover above the ground, eliminating friction and so cutting journey times. Children will play on hover bikes and hover boards.

The key to time travel will be uncovered allowing people to travel through wormholes (which bridge different places and times in the universe) and reappear in the future or the past.

New sea-craft will enable us to explore previously hidden ocean depths.

Teleportation pods will allow us to beam ourselves across the world.

WHAT SCIENCE SAYS

Transport An amphibious rescue vehicle has been invented by a Dutch company. Electric, gas and alcohol-based cars already exist and, in 1996, a British inventor launched a sugar-powered rocket. Solar-power conversion cells (that change solar energy to fuel) are currently very expensive but the price is falling.

Antimatter Research is underway and scientists can now produce one ten-billionth of a gram of antimatter! (In 1940 scientists could only make a few millionths of a gram of plutonium, but by 1945 they had a 10-kilo atomic bomb!)

Laser Tests have begun using lasers as rocket launchers. One 50-gram object reached speeds of almost Mach 5 (5 times the speed of sound).

Space holidays A re-usable, pilotless rocket starts testing in 2000. It needs a small ground crew and does not shed any parts during flights, making cheap space tourism a reality.

Space communities NASA has scheduled a Mars mission for 2020. The essential first tasks will be locating and drilling for water and recycling waste to produce oxygen.

Aircraft Pilotless planes are currently being developed such as *Dark Star*, a US spy plane. Russian engineers are building a 'flying saucer' equipped to carry 400 passengers. NASA plans to be flying solar planes by 2002.

Into orbit Planes that go into orbit such as the *Pathfinder* are being tested and by 2003 could be carrying up to 40 passengers. Travel agencies are now taking bookings and offering training in weightlessness.

Hover technology Magnetic-levitation (maglev) trains should be running between Berlin and Hamburg by 2005. In 1997, Japan's maglev train reached 401 km/h.

Sea exploration Current submarines are made of light materials, weighted to make them sink. Designer Heinz Lipschutz believes future craft will be made of heavy, strong materials, have wings and powerful engines, and reach even greater depths.

Teleportation In 1996, a 15-cm plastic model was scanned into a computer then 'beamed' across California, USA, to materialize as a 3D copy. However, actual teleportation still looks unlikely. Today's computers could not handle the data required to teleport human beings and one small error would result in a different creature being created!

Time travel
Wormholes are highly radio-active and likely to collapse. Until these dangers are overcome they are dangerous for time/space travel. Many physicists think time travel is forbidden by nature as it creates too many impossibilities.

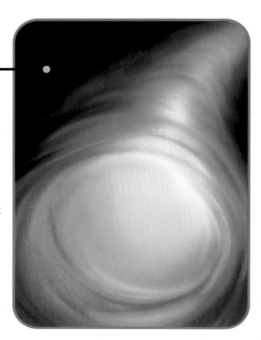

What Do You Think?

Could time travel serve a useful purpose?

Where would you travel to in time?

What will happen if the Moon becomes a popular holiday destination?

What problems might human teleportation cause?

If antimatter is so dangerous, should it be used?

A Healthy Future

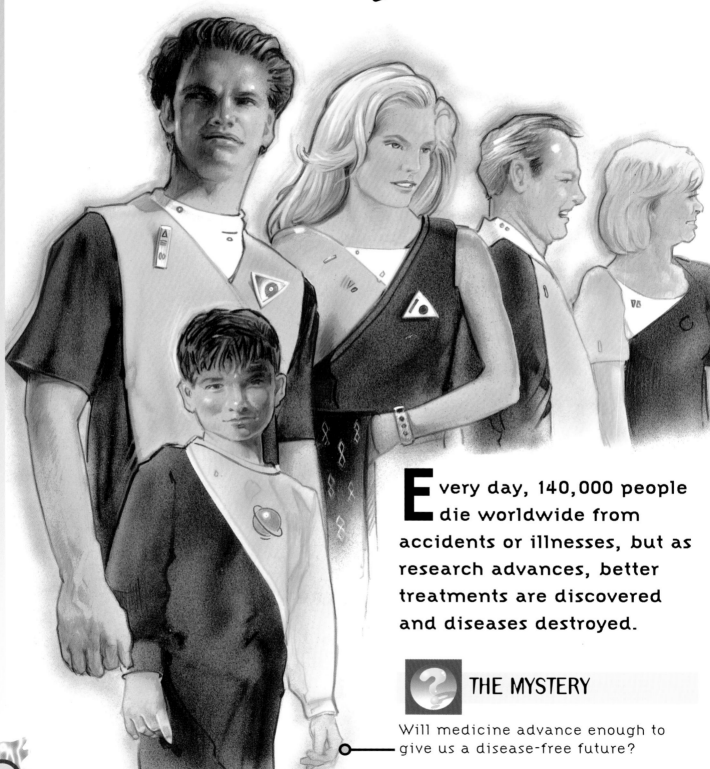

Every day, 140,000 people die worldwide from accidents or illnesses, but as research advances, better treatments are discovered and diseases destroyed.

THE MYSTERY

Will medicine advance enough to give us a disease-free future?

THE PREDICTIONS

Diagnosing illness will be easier. Wearable micro-computers will record data about our bodies. By plugging into a computer we will get advice and diagnosis on line. The results of home blood

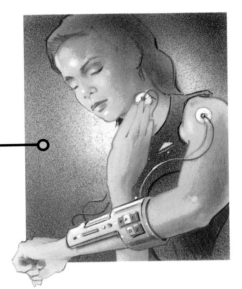

and urine tests will be e-mailed to a medical centre and the correct medication posted back.

Robots will prepare medicines, conduct experiments and even carry out operations. Micro-robots will be inserted into the body to carry out repairs and locate diseases.

Treatments will be more effective. A swallowed capsule will be tracked through the body until it reaches the infected area. A remote-control device will open the capsule and deliver medication to the exact spot.

Doctors will be trained using artificial human bodies and virtual surgery.

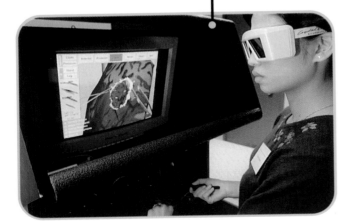

Serious conditions such as cancer, cystic fibrosis, asthma and diabetes will be eliminated as the individual genes responsible for them are identified. People likely to develop such illnesses will simply be inoculated against them.

A broken bone will be fixed quickly with an injection of liquid bone cement.

A simple injection will prevent tooth decay and ensure no more pain at the dentist.

All parts of the body will be replaceable including the brain (though this might result in a completely different personality in the same body). A supply of animal or artificial organs will be frozen and stored for future use.

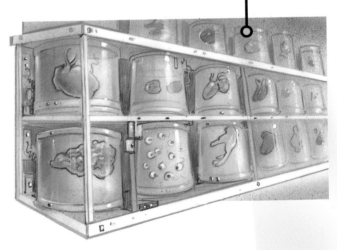

But new drug-resistant diseases will ravage the planet. Superbugs will wipe out thousands of people as scientists struggle to manufacture antidotes. Previously harmless germs could mutate into killers. As global warming heats up the planet, disease-carrying mosquitoes will spread.

However, natural alternative medicine will provide cures for diseases and supergerms that traditional medicine cannot fight.

Microscopic nano-particles which can be injected into the bloodstream to diagnose disease, are under development and in 1995, the California Institute of Technology engineered a micro-robot that was navigated through a pig's intestines.

Remote-control drugs Drugs that target only the sick (instead of all) cells are being developed for cancer treatments.

Transplants
Every organ except the brain and spinal cord can be replaced. The first voice transplant was done in 1998. At present, freezing causes damage to donor organs, but the recent discovery of an 'anti-freeze' solution means they could be safely stored.

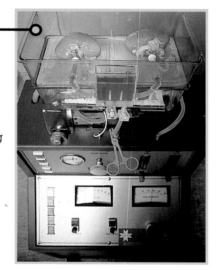

Broken bones French and American experiments with 'liquid bone' show that it sets as hard as normal bone in an hour. In France, Sweden, the Netherlands and USA, tests are currently being done to mend hip, wrist and knee fractures.

WHAT SCIENCE SAYS

Hi-tech medicine Tiny body monitors have been used successfully by climbers and astronauts. In 1994 a satellite 'telemedicine' link connected doctors in the Middle East and USA. Drug companies already use robots and the Americans are developing a remote-controlled robosurgeon.

Fighting disease Identifying the genes which cause diseases has already given hope to cystic fibrosis sufferers and diabetics and may be used in future cancer treatments.

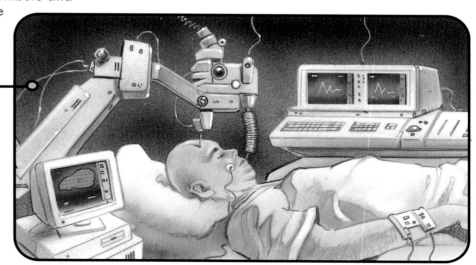

Teeth A vaccine is being tested which makes the body resistant to the bacteria that cause tooth decay.

New killers An antibiotic-resistant strain of a normally harmless germ killed an American recently. Some genetically modified foods contain an antibiotic-resistant gene. If this enters the human food chain, patients may not respond to antibiotic drugs. The appearance of AIDS and new types of killer flu (such as Asian Bird Flu in 1997) suggest we will never eliminate sickness. Old diseases, such as cholera and tuberculosis, are reappearing.

The World Health Organization predicts that heart-disease, strokes and cancer will still be the leading causes of death in 2025, so a disease-free future is a long way off.

Alternative Health Natural medicines such as acupuncture and herbal remedies are growing in popularity. Today, alternative health practitioners outnumber traditional GPs.

Mosquito diseases By 2025 the proportion of the world at risk from malaria could rise to 60%.

What Do You Think?

Is switching on a computer doctor better than going to your local surgery?

Would you want a robot to operate on you?

Should doctors practise on computer simulators before doing real operations?

Is alternative medicine a step backwards?

Without illness, what would happen to the world's population?

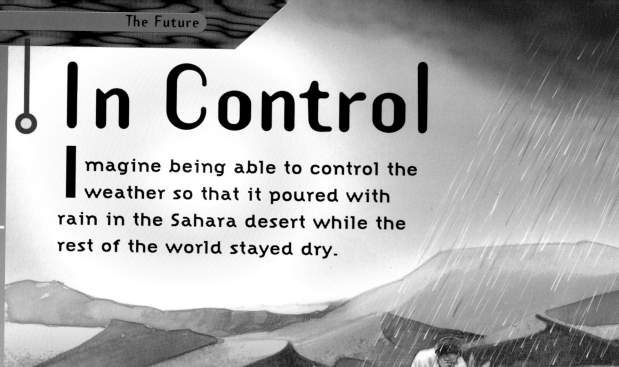

In Control

Imagine being able to control the weather so that it poured with rain in the Sahara desert while the rest of the world stayed dry.

THE MYSTERY

How far will we control our lives in the future?

THE PREDICTIONS

We will **eliminate famine** through genetically engineered crops. Genetically engineered cows will give more milk and animals will be cloned for cheaper meat.

The forces of nature such as the weather, volcanoes and earthquakes, will be under our control.

We will choose at what times programmes are shown on our TV sets. Interactive TV will give us the power to change the ending of our favourite movie every time we watch it. By plugging into a virtual reality control unit, we will control the camera angles, storyline and even take part.

It will be possible to instantly change the colour of our walls or the virtual view from our windows by the flick of a switch. ——

Politics will come under our direct influence. The views of everyone will be transmitted electronically through a computer whenever an important government decision needs to be taken.

WHAT SCIENCE SAYS

Food Scottish scientists are now able to produce hundreds of genetically identical sheep from cells in a laboratory dish. Genetically modified crops continue to be developed.

Nature Hurricanes, floods, droughts and earthquakes kill thousands. Scientists continue to look for reliable ways of predicting disasters. But prediction is far from control.

TV 'Video on Demand' is being tested around the world. People specify which programme they want to see and it is delivered down telephone or cable TV networks. Interactive TV may be available within 30 years.

Smart Homes A paint that changes colour via a remote control is now being developed.

Politics Electronic voting could soon be a reality. This technology could be used for getting people's views on important issues.

What Do You Think?

How much control do we need over politicians?

What movie would you change if you could?

Is controlling the weather a good thing?

If you were in a bad mood, what colour would you want the walls to be?

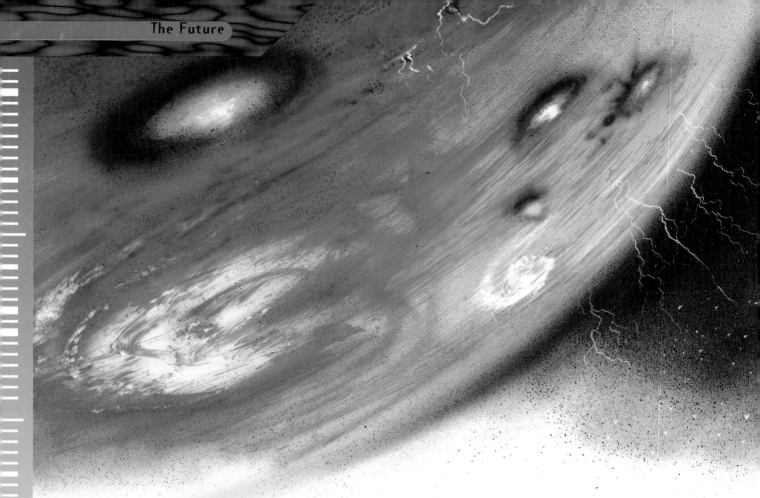

Saving the Planet

In the name of progress we destroy rainforests, endanger animal species, and choke our atmosphere with pollution.

 ## THE MYSTERY

Will future developments save or destroy the planet?

 ## THE PREDICTIONS

- **Things will get worse** before they get better. As technology advances, outdated computers and gadgets will pile up.

- **Pollution** will soar before car manufacturers develop cleaner engines. The world will heat up as we burn fossil fuels. The hole in the ozone layer will grow.

The polar ice-caps will melt and many areas will flood. The planet will become more unstable with dormant volcanoes erupting.

But technology will eventually save the planet. Factories will be replaced by sealed 'nanofactories'. Inside, microscopic machines will produce microscopic parts as gadgets become smaller. Nothing will be wasted and when a product is outdated it can be re-formed into something else.

Domestic nano-recyclers will break down household waste. As more people work from home, road congestion and pollution will ease.

A new, clean energy supply will be developed from helium and hydrogen (taken from water) to replace fossil fuels. Solar power will also be a planet-friendly alternative. Global warming will cease.

What Do You Think?

Will you want a new computer in five years?

If water can be used for fuel, what would happen to supplies?

Can governments do more to stop pollution?

Is everything recyclable?

WHAT SCIENCE SAYS

Waste technology A computer has a 5-year lifespan. In Germany, only 9% of electrical items are recycled.

In Moscow's Gorky Park, a disused space shuttle is now part of a children's playground.

Pollution 20% of the Western world's energy is consumed by transport, contributing to 20% of all greenhouse gases. There are already enough CFCs in the atmosphere to continue destroying the ozone layer for another 75 years.

Extreme weather is occurring around the globe. And in the mid 1980s, volcanoes in Papua New Guinea, California and Italy all became active.

Waste free Using nanomachines, waste products could be constantly recycled.

Safe energy Once scientists can produce enough commercially, hydrogen and helium could provide an energy source for the next billion years.

Solar power As the cost of making solar conversion cells falls, solar power could take off. In 1997, demand increased by 43%.

Extinct species China's national symbol, the panda, is to be cloned to save it from extinction.

Technology Take-over

In the 1980s, a US professor predicted self-flying aeroplanes with a crew of two: a pilot and a dog. The pilot would feed the dog; the dog would bite the pilot if he touched the controls.

? THE MYSTERY

Will technology reach a point where it is controlling us rather than the other way around?

THE PREDICTIONS

By 2025 machines will be smarter than humans. People will become wholly dependent on them to solve problems and make decisions.

With more choices to make, people will need electronic selectors. With thousands of TV programmes, an electronic guide will pick out items that it 'thinks' will interest you.

Machines will stop people breaking the law. The car computer will prevent the driver from taking an illegal short cut. 'Smart CDs' will refuse to make unauthorized recordings. Sensors around buildings will instantly inform the police of suspicious activity, recording acts of vandalism and identifying criminals through their ID implants.

Household gadgets will know what is 'best'. Your music system will select music to wake you up or relax you. Your automatic food maker will stop you guzzling fatty food, dishing out healthy salads instead.

WHAT SCIENCE SAYS

Smart Machines? *Deep Blue*, the mighty chess computer, managed just one win over champion Gary Kasparov in 1996, despite being able to 'consider' 250 million moves a second. However, chess-playing computers of the 21st century will be able to beat any human being.

Thinking TV Currently, search engines find information for us on the Internet. In Massachussetts, USA, an electronic TV guide has been developed which learns your viewing tastes and then browses on its own.

Law-abiding machines 'Smart paper' which refuses to make illegal copies, has been invented. In Singapore, urine sensors in lifts prevent 'anti-social' behaviour!

Keeping track Badges worn by staff in one company track their movements so that phone calls can be forwarded wherever they are.

Home knows best A Korean company have developed a 'thinking' music system which calculates your moods from your birthdate and then recommends suitable music.

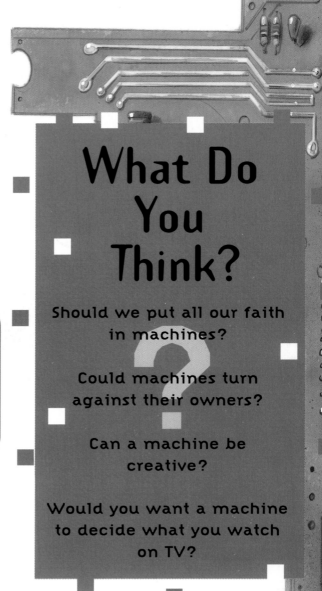

What Do You Think?

Should we put all our faith in machines?

Could machines turn against their owners?

Can a machine be creative?

Would you want a machine to decide what you watch on TV?

Humans

The human body changes very slowly, from generation to generation. But as science advances, these changes are not always due to evolution.

THE MYSTERY

With advances in science and medicine, will humans change beyond recognition?

THE PREDICTIONS

Many people are currently overweight but diet pills will soon allow us to effortlessly maintain our ideal body size, yet eat anything.

Different nationalities will look similar. The popularity of western junk food in China and Japan, for example, is already making people there taller with longer faces.

Men and women will dress identically. Clothes will have built-in air conditioners and heating panels, and shoes will massage aching feet. Fabrics will be waterproof, bacteria and radiation proof.

Non-human genes will be injected into the body to give us animal powers. We could run like a cheetah, have the acute hearing of a bat, or even grow wings.

Cyborgs (human robots) will become a reality. In 1994, scientists gave a blind boy some of his sight back by implanting a light-sensitive chip into his optic nerve. In the future, artificial parts with super-human capabilities will be implanted: a nose with a heightened sense of smell or super-strong bionic limbs.

Memory chips and electronic neurons will be inserted into our brains enabling us to think a million times faster. Pills that increase intelligence and learning will be available.

We will develop our psychic abilitites to operate computers by thought and to communicate telepathically.

'Old age' will begin much later, perhaps at 110 or 120. Ageing will be staved off by injecting minuscule 'nano-particles' to repair cell damage from inside the body.

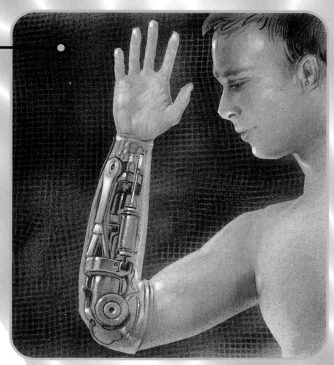

Parents will 'design' their own babies, choosing sex, eye-colour, hair type and personality.

Clones (exact replicas) of people will be grown in laboratories.

WHAT SCIENCE SAYS

Clothes The 1996 Olympics saw many new fabrics worn by athletes to improve performance and limit fatigue. One new material shields against electromagnetic waves, and a jacket with solar-heating panels and a built-in computer has been designed.

Human dimensions Slimming pills are already on the market and, as experiments with mice show, by controlling weight-reducing hormones in the body, genetic engineers hope to prevent obesity in future generations.

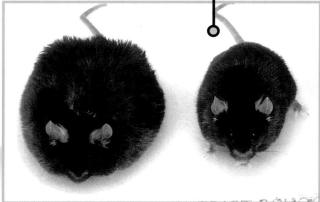

Transplants Baboon livers were transplanted into two men who both died within 3 months. In the next few years, scientists are hoping to successfully use pigs' organs for transplants. Scientists in Cambridge, UK, mixed human and crocodile blood to produce a blood type which could enable humans to stay under water longer.

Cyborgs Some scientists believe cyborgs to be the next stage in human evolution, but making a replica of the complex human brain is extremely difficult. Scientists in Germany have managed to connect a living leech nerve cell to a silicon chip so that the two can electrically interact.

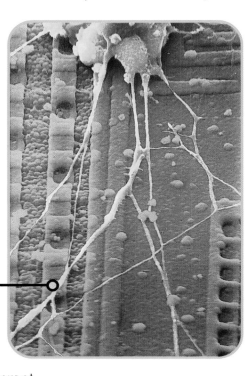

Human cloning Skin and a human ear have already been grown in a laboratory. In 1998, US physicist Richard Seed announced plans to clone himself.

Brain power American scientists have been working on thought-controlled planes since the late 1980s and have had some limited success on simulators. A cosmonaut on the *Mir* space station experimented with telepathy as a means of deep-space communication.

Old age British genetecist Steve Jones doubts whether the ageing process will ever be stopped because of the difficulty in isolating and shutting down every relevant gene.

Genetic engineering Our DNA pattern decides our characteristics. Scientists currently mapping the DNA in just one human cell have so far only uncovered 3% of the total, so 'designer babies' are still a remote possibility.

What Do You Think?

Would you like to live forever?

If one part of your body could be bionic, which would it be?

What jobs could cyborgs do?

Would you like to meet a clone of yourself?

Should scientists interfere with the human body?

Glossary

Alternative energy Fuels that do not harm the environment or use up the Earth's resources, eg. solar, wind and water power.

Alternative medicine Natural treatments such as acupuncture or homeopathy that are outside conventional western medicine.

Antibiotics Conventional medicine used to treat illnesses caused by bacteria.

Antimatter Particles of matter smaller than atoms. When matter meets antimatter a high-energy explosion results.

Atomic bomb A weapon of mass destruction developed by scientists in the 1940s and composed of minute particles of matter (atoms) which react together to produce a powerful explosion.

Cloning The creation of a copy of a cell taken from a plant or animal, which develops into a replica of the plant or animal.

Cyborg A creature that is part human, part robot.

DNA (deoxyribonucleic acid) The material in all living things that contains genetic information.

Donor organ A body part used for transplant, eg. heart, kidney, lung, liver.

Electronic book A book you read on a machine rather than on paper.

Electronic money A single 'smart card' that replaces cash or a credit card.

Evolution The gradual changing of a living thing from one form into another.

Extinct An extinct animal has died out.

Futurologist A person who forecasts the future by studying trends in technology.

Gated community A private, high-security estate with (usually) very wealthy residents.

Genes In a living body, the units inherited from the previous generation that pass on individual features, eg. eye colour.

Genetic engineering Altering genes to somehow change a living thing.

Genetically modified crops Food plants whose genes have been altered.

Global warming Changes in temperature and weather patterns caused by the hole in the ozone layer of the Earth's atmosphere.

Interactive TV Future television where viewers can directly influence the programmes they see.

Laser launchers Powerful light beams used to launch rockets.

Mach A measure of speed based on the speed of sound, eg. Mach 5 = 5 x the speed of sound.

Maglev (magnetic levitation) A method of transport via a high-speed hover-vehicle.

Nanotechnology The branch of technology that deals with manipulating atoms.

NASA (National Aeronautics and Space Administration) A US space research organization.

Neuron A nerve cell.

Psychic abilities Abilities channelled through the brain or mind, such as telepathy.

'Smart' Any object that can adapt itself to suit different circumstances.

Solar conversion cell A cell which transforms sunlight into a usable energy form.

Subatomic particle A particle of matter which is smaller than an atom, but contains an equal amount of energy.

Superbugs New strains of disease-causing bacteria and viruses that develop in response to antibiotics and vaccines.

Teleportation Transporting a person or object from one place to another in a non-material form, eg. as particles of matter.

Transplant To replace an unhealthy human organ with a healthy one.

Virtual reality An image or environment created by computer with which a person can interact as if it were real.

Waste technology The recycling of outdated and waste products.

World Health Organization The international organization that monitors disease, providing information and aid worldwide.

Index

600

This book is due for return on or before the last date shown below.

22/11/00		
06 FEB		
18/02/01		
11 FEB 2002		

Don Gresswell Ltd., London, N21 Cat. No. 1207

DG 02242/71

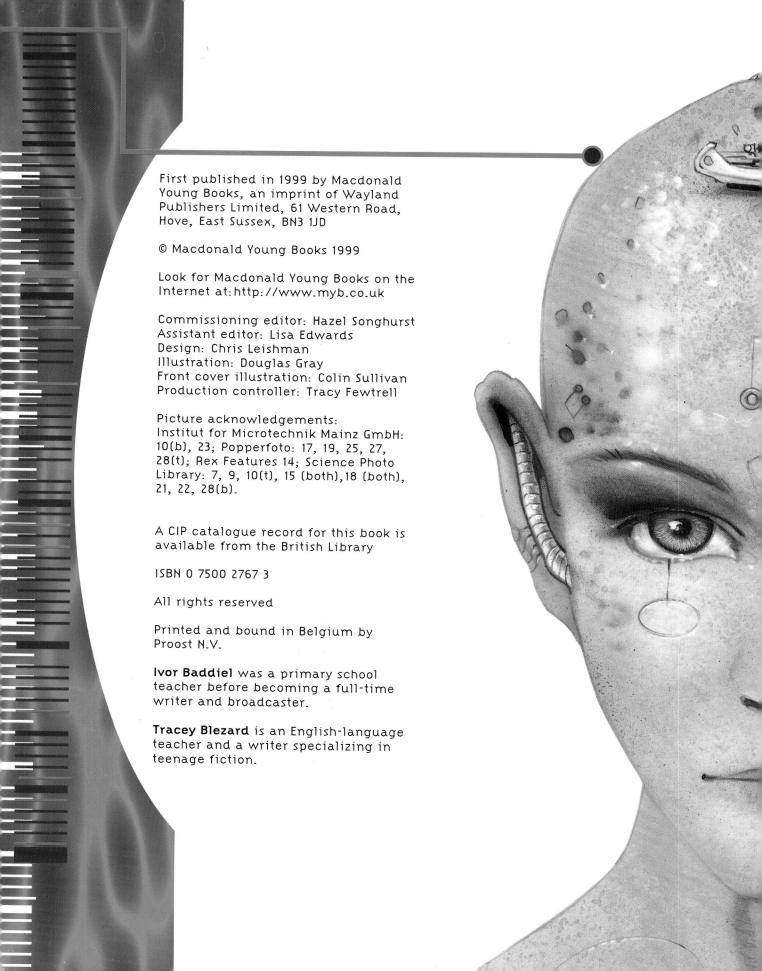

First published in 1999 by Macdonald
Young Books, an imprint of Wayland
Publishers Limited, 61 Western Road,
Hove, East Sussex, BN3 1JD

© Macdonald Young Books 1999

Look for Macdonald Young Books on the
Internet at: http://www.myb.co.uk

Commissioning editor: Hazel Songhurst
Assistant editor: Lisa Edwards
Design: Chris Leishman
Illustration: Douglas Gray
Front cover illustration: Colin Sullivan
Production controller: Tracy Fewtrell

Picture acknowledgements:
Institut for Microtechnik Mainz GmbH:
10(b), 23; Popperfoto: 17, 19, 25, 27,
28(t); Rex Features 14; Science Photo
Library: 7, 9, 10(t), 15 (both),18 (both),
21, 22, 28(b).

A CIP catalogue record for this book is
available from the British Library

ISBN 0 7500 2767 3

Printed and bound in Belgium by
Proost N.V.

Ivor Baddiel was a primary school
teacher before becoming a full-time
writer and broadcaster.

Tracey Blezard is an English-language
teacher and a writer specializing in
teenage fiction.